IPCS

Health and Safety Guide No. 56

HYDRAZINE
HEALTH AND SAFETY GUIDE

This is a companion volume to
Environmental Health Criteria 68: Hydrazine

Published by the World Health Organization for the International Programme on Chemical Safety (a collaborative programme of the United Nations Environment Programme, the International Labour Organisation, and the World Health Organization)

WORLD HEALTH ORGANIZATION, GENEVA 1991

This report contains the collective views of an international group of experts and does not necessarily represent the decisions or the stated policy of the United Nations Environment Programme, the International Labour Organisation, or the World Health Organization

WHO Library Cataloguing in Publication Data

Hydrazine : health and safety guide.

(Health and safety guide ; no. 56)

1.Hydrazines - standards I.Series

ISBN 92 4 151056 0 (NLM Classification: QU 60)
ISSN 0259-7268

The designations employed and the presentation of the material in this publication do not imply the expression of any opinion whatsoever on the part of the Secretariat of the World Health Organization concerning the legal status of any country, territory, city or area or of its authorities, or concerning the delimitation of its frontiers or boundaries.

The mention of specific companies or of certain manufacturers' products does not imply that they are endorsed or recommended by the World Health Organization in preference to others of a similar nature that are not mentioned. Errors and omissions excepted, the names of proprietary products are distinguished by initial capital letters.

Computer typesetting by HEADS, Oxford OX7 2NY, England
Printed by Wissenschaftliche Verlagsgesellschaft mbH · D-7000 Stuttgart 10

CONTENTS

CONTENTS

INTRODUCTION

The Environmental Health Criteria (EHC) documents produced by the International Programme on Chemical Safety include an assessment of the effects on the environment and on human health of exposure to a chemical or combination of chemicals, or physical or biological agents. They also provide guidelines for setting exposure limits.

The purpose of a Health and Safety Guide is to facilitate the application of these guidelines in national chemical safety programmes. The first three sections of a Health and Safety Guide highlight the relevant technical information in the corresponding EHC. Section 4 includes advice on preventive and protective measures and emergency action; health workers should be thoroughly familiar with the medical information to ensure that they can act efficiently in an emergency. Within the Guide is a Summary of Chemical Safety Information which should be readily available, and should be clearly explained, to all who could come into contact with the chemical. The section on regulatory information has been extracted from the legal file of the International Register of Potentially Toxic Chemicals (IRPTC) and from other United Nations sources.

The target readership includes occupational health services, those in ministries, governmental agencies, industry, and trade unions who are involved in the safe use of chemicals and the avoidance of environmental health hazards, and those wanting more information on this topic. An attempt has been made to use only terms that will be familiar to the intended user. However, sections 1 and 2 inevitably contain some technical terms. A bibliography has been included for readers who require further background information.

Revision of the information in this Guide will take place in due course, and the eventual aim is to use standardized terminology. Comments on any difficulties encountered in using the Guide would be very helpful and should be addressed to:

The Manager
International Programme on Chemical Safety
Division of Environmental Health
World Health Organization
1211 Geneva 27
Switzerland

THE INFORMATION IN THIS GUIDE SHOULD BE CONSIDERED AS A STARTING POINT TO A COMPREHENSIVE HEALTH AND SAFETY PROGRAMME

1. PRODUCT IDENTITY AND USES

1.1 Identity

Common name:	hydrazine
Chemical formula:	N_2H_4
Chemical structure:	$H_2N{-}NH_2$
Common synonyms:	diamide, diamine, anhydrous hydrazine, hydrazine base
Common trade names (of mixtures):	Aerozine-50 (a 1:1 w/w fuel mixture of anhydrous hydrazine and 1,1-dimethylhydrazine); Hydrazine hydrate ($N_2H_4.H_2O$) (a 1:1 molar mixture of anhydrous hydrazine and water); Levoxin (a 15–64% aqueous solution); SCAV-OX (a 35–64% aqueous solution); Zerox (a 15–64% aqueous solution)
CAS registry number:	302-01-2
RTECS registry number:	MU7175000
UN number:	2029 (2030 for hydrazine monohydrate $N_2H_4.H_2O$)
Conversion factors:	1 ppm = 1.31 mg/m^3 at 25 °C and 101.3 kPa (760 mmHg) 1 mg/m^3 = 0.76 ppm

1.2 Physical and Chemical Properties

Anhydrous hydrazine, and aqueous solutions, are colourless liquids. Anhydrous hydrazine is a caustic, fuming, hygroscopic liquid at normal temperature and pressure. It has an ammoniacal, fishy, and pungent odour. The odour perception threshold is 4–9 mg/m^3. A 1:1 mixture of anhydrous hydrazine in water (hydrazine monohydrate) containing 64%

by weight of hydrazine, fumes slightly in air and has an ammoniacal odour. Hydrazine is infinitely soluble in water and may be flammable and explosive up to a concentration of 400 g/litre (40%). Hydrazine is basic, and is a strong reducing agent. Some physical and chemical properties of hydrazine are given in the Summary of Chemical Safety Information (section 6).

1.3 Composition

Hydrazine is often sold as aqueous hydrazine (15–64% solutions of anhydrous hydrazine in water) or as hydrazine hydrate.

1.4 Production and Uses

The total world production capacity for hydrazine was estimated to be in excess of 35 000 tonnes in 1981. Significant production capacities are reported for the Federal Republic of Germany, France, Japan, the United Kingdom, and the USA. Hydrazine is mainly used as a raw material in the manufacture of agricultural chemicals, blowing agents, polymerization catalysts, and pharmaceutical products, and as a corrosion inhibitor in boiler water. Both the hydrate and anhydrous hydrazine are used as propellant fuels for aircraft and spacecraft.

2. SUMMARY AND EVALUATION

2.1 Human Exposure to Hydrazine

Hydrazine is not known to occur naturally, except, perhaps, in tobacco. It can be released into the atmosphere during venting operations, storage, or transfer. The total emission is estimated to represent nearly 0.01% of the hydrazine produced. Accidental discharges into water, air, and soil can result from bulk storage, handling, transport, or improper waste disposal.

Exposure of human beings to hydrazine may occur occupationally or accidentally, from hydrazine-based drugs, or from the use of tobacco. It has been shown that concentrations of up to 0.35 mg/m^3 can occur during production under normal conditions, and that, exceptionally, concentrations of up to 1.18 mg/m^3 may occur. During the handling of the fuel, concentrations of up to 0.25 mg/m^3 have been measured under normal conditions and, exceptionally, up to 2.59 mg/m^3. Because the compound is degraded so rapidly in the environment (section 2.2), measurable levels are not normally encountered, and, therefore, hydrazine does not pose a significant hazard for the general population. However, the general population may be exposed to hydrazine vapour through accidental discharge. Evaporation of hydrazine from a liquid spill can be sufficient to generate an atmospheric concentration as high as 4 mg/m^3.

2.2 Fate of Hydrazine

Hydrazine is degraded rapidly in the air, through reactions with ozone, hydroxyl radicals, or nitrogen dioxide. In polluted air, the life-time will be approximately 1 h. In soil, aqueous hydrazine is adsorbed and decomposed on clay surfaces, under aerobic conditions. However, available data are inadequate to describe the behaviour of hydrazine in the soil. The degradation rate of hydrazine in water is highly dependent on various factors, such as pH, temperature, oxygen content, alkalinity, hardness, and the presence of organic material and metallic ions. Hydrazine is degraded rapidly under aerobic conditions in the presence of organic material, and/or in alkaline or hard water. It is more persistent in soft, metal-free water. Hydrazine is biodegradable by microorganisms in activated sludge. However, at concentrations above 1 mg/litre, hydrazine is also toxic for these microorganisms, especially for nitrifying bacteria. Hydrazine does not bioaccumulate.

2.3 Effects on Organisms in the Environment

The concentration of hydrazine that is lethal for half the number of fish in a population (LC_{50}) exposed for 1–4 days, ranged from 0.54 to 5.98 mg/litre. The lowest-observed-effect level, found in a fathead minnow embryo-larvae test, was 0.1 mg/litre. Nitrifying bacteria in activated sludge are inhibited at levels higher than 1 mg/litre. Many microorganisms are more sensitive, and show threshold levels as low as 0.00008 mg/litre, reported for the blue alga *Microcystis aeroginosa*.

Hydrazine can inhibit germination in plants, and is toxic for plants in both air and water.

On the basis of these data, it can be concluded that hydrazine may present a hazard for aquatic organisms and plant life.

2.4 Effects on Animals

Hydrazine is rapidly absorbed through the skin or via other routes of exposure. It is also rapidly distributed to, and eliminated from, most tissues. In mice and rats, the absorbed hydrazine is excreted via the urine, partly unchanged, and partly as labile conjugates or as acid-hydrolysable derivatives. Metabolism of hydrazine produces a significant amount of nitrogen, which is excreted via thc lungs.

Single oral doses of 55–64 mg/kg body weight and vapour concentrations of 750 mg/m^3 for 4 h were lethal for half the number of exposed rats in a population (LD_{50}, LC_{50}). Thus, the compound is moderately toxic according to the scale of Hodge & Sterner.

Most of the effects on human beings (section 2.5) exposed to hydrazine have also been observed in experimental animals. In addition, loss of body weight, anaemia, hypoglycaemia, fatty liver, and convulsions have been frequently reported. Fatty liver was reported in mice, and body weight loss in rats, when they were exposed continuously for 6 months to 0.26 mg/m^3, the lowest of three exposure levels; monkeys and dogs were not affected at this concentration. There are no data on which to establish a no-observed-effect level for the inhalation route, but, in a 7-month drinking-water study, a no-observed-effect level of 3 µg/kg body weight was reported.

Studies on rats and mice have indicated that hydrazine produces adverse effects on embryos and fetuses, when administered at doses that are toxic for the mother. The adverse effects include increased resorptions, reduced fetal weight, increased perinatal mortality, and increased incidences of litters and fetuses with abnormalities. The abnormalities include primarily supernumerary and fused ribs, delayed ossification, moderate hydronephrosis, and moderate dilation of the brain ventricle.

Hydrazine induced gene mutations and chromosome aberrations in a variety of test systems, including plants, phages, bacteria, fungi, *Drosophila*, and mammalian cells *in vitro*. Indirect alkylation was introduced in liver DNA of rodents after *in vivo* exposure to toxic doses. Hydrazine also caused DNA damage *in vitro*. It transformed hamster and human cells *in vitro*, but did not increase unscheduled DNA synthesis in the germ cells of mice *in vivo*, or induce chromosome aberrations, micronuclei, or dominant lethals in mice *in vivo*. In rats, it was reported that hydrazine induced chromosome aberrations *in vivo*.

Hydrazine vapour induced nasal tumours, most of which were benign, in F-344 rats and Syrian golden hamsters, but not in C57BL/6 mice, after 12 months of treatment and life-time observation. In several limited gavage and drinking-water studies, hydrazine induced an increased incidence, in some cases dose-related, of multiple pulmonary tumours in various mouse strains and Cb/Se rats. In two strains of mice, an increased incidence of hepatocarcinomas was also induced. A very low, but increased, incidence of hepatocarcinomas was observed in male Cb/Se rats. No tumours were observed in orally-exposed hamsters.

On the basis of the carcinogenicity studies on experimental animals, there is evidence that hydrazine is an animal carcinogen.

2.5 Effects on Human Beings

In cases of acute human poisoning, vomiting, severe irritation of the respiratory tract with pulmonary oedema, central nervous system depression, and hepatic and renal damage have been reported. No data are available from which to make an estimate of the level of hydrazine inhaled in cases of acute poisoning by the respiratory route. However, from reports of poisoning by the oral route, it would appear that ingestion of amounts of the order of 20–50 ml would cause severe intoxication and may be lethal. From the available human data, it is not possible to estimate a no-observed-effect dose. Pyridoxine treatment of poisoned human beings has been reported to be successful on several occasions.

Skin and eye irritation have been observed in human beings who had come into contact with hydrazine, but the data are insufficient to establish a no-observed-effect level. Hydrazine is a strong skin sensitizer in human beings. Once sensitized to hydrazine, a person may also become sensitive to hydrazine derivatives (cross-sensitization).

Data are lacking concerning the effects of hydrazine on the human embryo or fetus. In the absence of human data, and on the basis of the animals studies, it is prudent to assume that hydrazine would have adverse effects on the human embryo or fetus at exposure levels near to those producing toxic effects in the mother. Such levels may occur from accidental spillages.

Data are inadequate to assess the carcinogenicity of hydrazine in human beings. However, taking into account the mutagenicity data, as well as the carcinogenicity data in animals, it would be prudent to consider hydrazine as a possible human carcinogen. Thus, exposure of human beings should be kept as low as feasible.

3. CONCLUSIONS

Hydrazine can be regarded as posing little hazard for the general population, under normal conditions. However, in the workplace, and under conditions of accidental exposure, hydrazine can present a significant health hazard. Human data are limited, but show that hydrazine may affect the central nervous system, liver, and kidneys. In addition, it may produce skin and eye irritation and skin sensitization. The results of animal studies suggest that effects on human beings may also include embryotoxicity, at levels near those producing toxic effects in the mother, and adverse effects on the respiratory system. On the basis of the evidence of carcinogenicity in animals, it would be prudent to consider hydrazine as a possible human carcinogen. Regarding the effects on the environment, it can be concluded that hydrazine may present a hazard for both aquatic organisms and plant life.

4. HUMAN HEALTH HAZARDS, PREVENTION AND PROTECTION, EMERGENCY ACTION

4.1 Main Human Health Hazards, Prevention and Protection, First Aid

The main human health hazards associated with certain types of exposure to hydrazine, together with preventive and protective measures and first-aid recommendations, are listed in the Summary of Chemical Safety Information (section 6).

4.2 Advice to Physicians

Pyridoxine treatment has been claimed to be successful on several occasions and may be considered. However, one case report mentions a reversible peripheral neuropathy resulting from treatment with a very large dose of pyridoxine. Otherwise, treat symptomatically. Pay attention to the possible development of pulmonary oedema and of damage to the liver. Skin burns should to be treated as for alkali or thermal burns.

4.3 Health Surveillance Advice

Human beings handling hydrazine, or hydrazine solutions, should undergo medical examination once a year, with emphasis on the functioning of the central nervous system, liver, and kidneys, and on disorders of the skin, eyes, and blood. It should be kept in mind that hydrazine is a possible human carcinogen and a strong skin sensitizer, and that cross-sensitization to hydrazine derivatives may occur.

4.4 Explosion and Fire Hazards

4.4.1 *Explosion hazards*

Anhydrous hydrazine–air mixtures, containing 4.7% or more (by volume) of hydrazine, may be explosive at temperatures higher than 38 °C, and can be ignited by sources of heat or ignition, or by ultraviolet radiation. The flash-point of hydrazine hydrate is 75 °C. The flammability of aqueous hydrazine decreases, and the flash-point increases, with increasing dilution, a 40% solution being just ignitable. Contact with metals, metal oxides, oxidizing substances, acids, and porous materials, such as earth, wood, asbestos, paper, or cloth, may cause fire and explosions. Contaminated clothing and equipment are fire hazards. Toxic decomposition products include nitrogen oxides and ammonia.

4.4.2 *Fire hazards*

Anhydrous hydrazine and aqueous hydrazine containing more than 40% of the compound are flammable liquids.

4.4.3 *Prevention*

For anhydrous hydrazine, use a closed system, where feasible, and explosion-proof electrical equipment. Do not use hydrazine in the vicinity of flames, sparks, and other sources of ignition. Do not smoke. Keep containers out of direct sunlight. Avoid contact between the compound and metals, metal oxides, acids, oxidizing agents, and porous materials. Prevent contamination of hydrazine. In case of fire, keep drums cool by spraying with water. Do not extinguish a fire caused by hydrazine, unless its release can be stopped. Fire-fighters need self-contained breathing apparatus, eye protection, and full protective clothing.

4.4.4 *Fire-extinguishing agents*

Water sprays, dry chemical, or carbon dioxide can be used on small fires. For large fires, alcohol foam, or water spraying to flood levels, can be used.

4.5 Storage

Hydrazine should be stored in tightly closed, well-labelled containers in an inert atmosphere, in a clean, well-ventilated area with controlled drainage. Store away from oxidizing agents, acids, metals, metal oxides, porous materials, direct sunlight, or other sources of heat or ignition. The containers should be connected to earth to prevent static sparks.

4.6 Transport

Current international regulations require that hydrazine hydrate, and aqueous solutions of it, should be transported in metal containers with polyethylene liners, in plastic canisters, or in stainless steel containers.

4.7 Spillage and Disposal

4.7.1 *Spillage*

Remove all ignition sources and evacuate the danger area. Collect leaking liquid in sealable containers. Keep spills from entering water sources and sewers. Promptly dilute spilled hydrazine with water spray, to produce at least a 40% solution. Construct barriers to contain the spill, or flush it into a container. Foam can be applied to retard vaporization. Use sand to collect small spills. Ensure personal protection, by using a self-contained breathing apparatus, eye protection, and full protective clothing.

4.7.2 *Disposal (based on the IRPTC waste disposal file)*

Hydrazine can be disposed of by dilution with water to produce at least a 40% solution and subsequent neutralization with dilute sulfuric acid. The resulting solution can be drained into the sewer, with abundant water.

Alternatively, hydrazine waste can be burnt in a chemical incinerator equipped with an after-burner and scrubber, after dilution with alcohol or another hydrocarbon fuel.

5. HAZARDS FOR THE ENVIRONMENT AND THEIR PREVENTION

Degradation of hydrazine in water may be slow, depending on conditions. Hydrazine may present a hazard for aquatic organisms and plant life.

Contamination of soil, water, and the atmosphere can be avoided by proper methods of storage, transport, and waste disposal. In case of spillage, apply methods recommended in section 4.7.1. When using exhaust ventilation, an exhaust scrubber may be needed. Use closed systems where feasible. Use dilute solutions of hydrazine rather than concentrated solutions, where possible.

6. SUMMARY OF CHEMICAL SAFETY INFORMATION

This summary should be easily available to all health workers concerned with, and users of, hydrazine. It should be displayed at, or near, entrances to areas where there is potential exposure to hydrazine, and on processing equipment and containers. The summary should be translated into the appropriate language(s). All persons potentially exposed to the chemical should also have the instructions in the summary clearly explained.

Space is available for insertion of the National Occupational Exposure Limit, the address and telephone number of the National Poison Control Centre, and for local trade names.

SUMMARY OF CHEMICAL SAFETY INFORMATION

HYDRAZINE

(diamide, diamine, hydrazine anhydrous, hydrazine base)

($H_2N–NH_2$)

PHYSICAL PROPERTIES	ANHYDROUS HYDRAZINE (100%)	HYDRAZINE HYDRATE (64%)	OTHER CHARACTERISTICS
Relative molecular mass	32.05		Colourless liquid, fuming in air, with an ammoniacal, pungent, fishy odour; aqueous solutions are colourless liquids; a 64% solution fumes slightly in air and has an ammoniacal odour; both hydrazine and an aqueous solution of hydrazine react strongly with acids and oxidizing agents; decomposition is accelerated by porous materials and some metals; toxic decomposition products are nitrogen oxides and ammonia; can have adverse effects well below the odour threshold
Melting point (°C)	2	-51.9	
Boiling point (°C)	113.5	120.1	
Water solubility	infinite	infinite	
Density (20 °C)	1.008 g/ml	1.032 g/ml	
Relative vapour density	1.1		
Vapour pressure (20 °C)	1.39 kPa	1 kPa	
Log *n*-octanol/water partition coefficient	-3.08		
Flammability (explosive) limits (%)	4.7-100	3.4-100	

HAZARDS/SYMPTOMS	PREVENTION AND PROTECTION	FIRST AID
SKIN: corrosion by liquid; burns; irritation by vapour; may enter body through skin	Wear clean, body-covering, impervious clothing and gloves; do not wear leather garments	Remove contaminated clothing under shower; rinse with plenty of water
EYES: corrosion by liquid; burns; irritation by vapour	Wear safety face-shield, or eye protection, in combination with breathing protection	Flush with plenty of water for at least 15 minutes
INHALATION: irritation of nose and respiratory tract; effects on central nervous system, liver, and kidneys	Apply explosion-proof ventilation and local exhaust, and, for non-routine activities, self-contained breathing apparatus	Remove victims to fresh air, and keep quiet; if breathing has stopped, apply artificial respiration; transport to hospital
INGESTION: corrosive effects; effects on central nervous system, liver, and kidneys	Do not eat, drink, or smoke during work	Rinse mouth; give plenty of water, milk, or lemon juice to drink; induce vomiting in conscious patients; transport to hospital

SUMMARY OF CHEMICAL SAFETY INFORMATION *(continued)*

HAZARDS/SYMPTOMS	PREVENTION AND PROTECTION	FIRST AID
GENERAL: the compound should be considered as a possible human carcinogen		
ENVIRONMENT: may be hazardous for aquatic and plant life	Apply proper methods of storage, transport, waste disposal, and handling of spills; when using exhaust ventilation, an exhaust scrubber may be needed; use closed systems and dilute solutions where feasible	

SPILLAGE	STORAGE	FIRE AND EXPLOSION
Remove ignition sources; evacuate area; collect leaking liquid in sealable container; promptly dilute spilled hydrazine with water spray; contain spills; ensure personal protection (use self-contained breathing apparatus, eye protection, and fully protective clothing)	Store in tightly-closed, well-labelled containers in an inert atmosphere, in a clean, well-ventilated area; store away from oxidizing agents, acids, metals, metal oxides, porous materials, and direct sunlight; connect containers to earth	Flammable; no open flames or other sources of ignition; vapour-air mixtures are explosive above 38 °C; use closed-system ventilation and explosion-proof electrical equipment; contaminated clothing and equipment are a fire hazard; in case of fire, keep drums cool by spraying with water; extinguish fires with spray or alcohol foam
WASTE DISPOSAL	**NATIONAL INFORMATION**	
Dilution with water and neutralization with dilute sulfuric acid; burning in a chemical incinerator equipped with an after-burner and scrubber	National Occupational Exposure Limit: National Poison Control Centre:	UN 2029 (100% hydrazine) UN 2030 (hydrazine monohydrate)

7. CURRENT REGULATIONS, GUIDELINES, AND STANDARDS

The information given in this section has been extracted from the International Register of Potentially Toxic Chemicals (IRPTC) legal file. A full reference to the original national document from which the information was extracted can be obtained from IRPTC. When no effective date appears in the IRPTC legal file, the year of the reference from which the data are taken is indicated by (r).

The reader should be aware that regulatory decisions about chemicals taken in a certain country can only be fully understood in the framework of the legislation of that country. The regulations and guidelines of all countries are subject to change and should always be verified with appropriate regulatory authorities before application.

7.1 Exposure Limit Values

Some exposure limit values are given in the table on pp. 26–27.

7.2 Specific Restrictions

The Joint FAO/WHO Meeting on Pesticide Residues has established an Acceptable Daily Intake (ADI) of 0–5 mg/kg body weight for maleic hydrazine containing not more than 1 mg hydrazine/kg maleic hydrazine (effective date: 1984).

In the USA, those registering technical maleic hydrazide under the pesticide registration regulations must submit a confidential statement of the formula to certify that the hydrazine level in their products does not exceed 15 mg/kg (effective date: 1983).

In the Federal Republic of Germany, the handling of hydrazine is prohibited or restricted for pre-adults and pregnant or nursing women (effective date: 1980).

CURRENT REGULATIONS, GUIDELINES, AND STANDARDS

7.3 Labelling, Packaging, and Transport

The European Economic Community regulations state that the label for >64% solutions of hydrazine should read as follows (effective date: 1982 (r)):

> *Very toxic by inhalation, in contact with skin and if swallowed; causes burns; possible risks of irreversible effects; wear suitable protective clothing, gloves and eye/face protection; in case of accident or if you feel unwell, seek medical advice immediately (show the label where possible).*

T+

The label is:

Meget giftig
Sehr giftig
Λίαν τοξικό
Very toxic
Très toxique
Molto tossico
Zeer vergiftig

For 15–64% solutions of hydrazine, the label should read (effective date: 1982 (r)):

> *Toxic in contact with skin and if swallowed; causes burns; in case of contact with eyes, rinse immediately with plenty of water and seek medical attention).*

The percentage concentration must be stated on the label. The label is:

C

* ES: Corrosivo
DA: Ætsende
DE: Ätzend
EL: Διαβρωτικό
EN: Corrosive
FR: Corrosif
IT: Corrosivo
NL: Corrosief
* PT: Corrosivo

CURRENT REGULATIONS, GUIDELINES, AND STANDARDS

EXPOSURE LIMIT VALUES

Medium	Specification	Country/ organization	Exposure limit description[a]	Value	Effective date
AIR	Occupational	Australia	Threshold limit value (TLV)		1985 (r)
			- Time-weighted average (TWA)	$0.1\ mg/m^3$ [b]	
		Czechoslovakia	Maximum allowable concentration (MAC)		1985
			- Time-weighted average (TWA)	$0.05\ mg/m^3$	
			- Ceiling value	$0.1\ mg/m^3$	1985
		Germany, Federal Republic of	Technical reference concentration		1986 (r)
			- 1-year time-weighted average	$0.13\ mg/m^3$ [b,c,d]	
		Italy	Threshold limit value (TLV)		1985 (r)
			- Time-weighted average (TWA)	$0.13\ mg/m^3$ [b,c]	
		Sweden	Threshold limit value (TLV)		1985
			- Time-weighted average (TWA)	$0.1\ mg/m^3$ [b,c,d]	
			- Short-term exposure limit (STEL) (15-min TWA)	$0.4\ mg/m^3$	

		USA (ACGIH)	Threshold limit value (TLV)		1985 (r)
			- Time-weighted average (TWA)	0.1 mg/m3 [b,c]	
		USA (OSHA)	Threshold limit value (TLV)		
			- Time-weighted average (TWA)	1 mg/m^3	
		USA (NIOSH)	Threshold limit value (TLV)		
			- Time-weighted average (TWA)	1 mg/m^3	
			- Ceiling value	2 mg/m^3	
		USSR	Ceiling value (including derivatives)	0.1 mg/m^3	1977
WATER	Surface	USSR	Maximum allowable concentration	0.01 mg/litre	1983

a TWA = time-weighted average over one working day (usually 8 h).
b Skin absorption.
c (Suspected of) carcinogenic (potential).
d Sensitization.

BIBLIOGRAPHY

FAO (1985a) *Guidelines for the packaging and storage of pesticides.* Rome, Food and Agriculture Organization of the United Nations.

FAO (1985b) *Guidelines for the disposal of waste pesticides and pesticide containers on the farm.* Rome, Food and Agriculture Organization of the United Nations.

FAO/WHO (1986) *Guide to Codex recommendations concerning pesticide residues.* Part 8. *Recommendations for methods of analysis of pesticide residues.* 3rd ed. Rome, Codex Committee on Pesticide Residues.

GIFAP (1982) *Guidelines for the safe handling of pesticides during their formulation, packing, storage, and transport.* Brussels, Groupement International des Associations Nationales des Fabricants de Produits Agrochimiques.

IARC (1972-present) *IARC monographs on the evaluation of carcinogenic risk of chemicals to man.* Lyon, International Agency for Research on Cancer.

IRPTC (1985) *IRPTC file on treatment and disposal methods for waste chemicals.* Geneva, International Register for Potentially Toxic Chemicals, United Nations Environment Programme.

IRPTC (1987) *IRPTC legal file 1986.* Geneva, International Register for Potentially Toxic Chemicals, United Nations Environment Programme.

PLESTINA, R. (1984) *Prevention, diagnosis, and treatment of insecticide poisoning.* Geneva, World Health Organization (Unpublished report No.VBC/84.889).

SAX, N.I. (1984) *Dangerous properties of industrial materials.* New York, Van Nostrand Reinhold Company, Inc.

UNITED NATIONS (1989) *Recommendations on the transport of dangerous goods.* 6th ed. New York, United Nations.